Introduction

In 2019, Carrie and her husband visited the Caymen Islands and were amused/entertained by the feral chickens running around in the streets. Upon returning home, Carrie and her cousin, Kate, began to wonder how a family might best take care of chickens during a storm. After a few conversations and a stack of sticky notes, Lola and the Hurricane was done.

There was a need for a book that helps children learn about evacuation preparation for a hurricane in an age-appropriate way. Lola's story shows little ones how being prepared for a natural disaster does not need to be scary.

We hope you enjoy it!
-Kate & Carrie

Published by CEY Press
821 Grand Ave, Suite 119
Pflugerville, TX 78660

ISBN:978-1-954885-18-9

LOLA & THE HURRICANE

BY CARRIE CASEY & KATE WOODWARD YOUNG

ILLUSTRATED BY SILVER STAR

"Grandma, why are you making so much food? Are we having a party?" asks Lola.

Grandma laughs. "No, Lola. A hurricane is coming."

"What is a hurricane?" asks Lola.

Grandma answers, "A hurricane is a big storm. A really big storm."

CHALK

“Hurricane? Is that like the rain that washed away my drawing? I don’t see why we need food for that!”

“It is a much bigger storm. It can knock things over and might break the power lines and flood the house. We are going to evacuate. We have to leave town and go away from the beach," Grandma answers.

“Let me tell you about a hurricane from when your daddy was little.
“Why?“

A big hurricane was coming to the island. We didn't know, so we didn't prepare or evacuate."

"Most folks didn't have weather radios and there was no alarm. When the storm hit it broke many of the houses, including the chicken houses."

"Is that why all the chickens run around in Georgetown?"

Yes, your Daddy lost his favorite chicken

"I don't want to leave! I like our house!"
I know, love, but you don't want to lose our chickens do you?
No, but..."

"As you prepare to evacuate, ensure that you have everything you need.

Take with you
food and water
for three days,
phones with
backup batteries,
illumination,
heat, pets,
medication,
something to keep
you entertained,
and changes of
clothing.

"Do we have all of that?"
"Let's go find out!"

Grandma asks, “What else do we need to take?”

Lola looks around and says, “How about some things we just love?”

My first
Owl Book

Lola stops her dad and asks, “What books should we bring?”

“Let’s go look,” says dad.

"I think we have what we
need from the house."

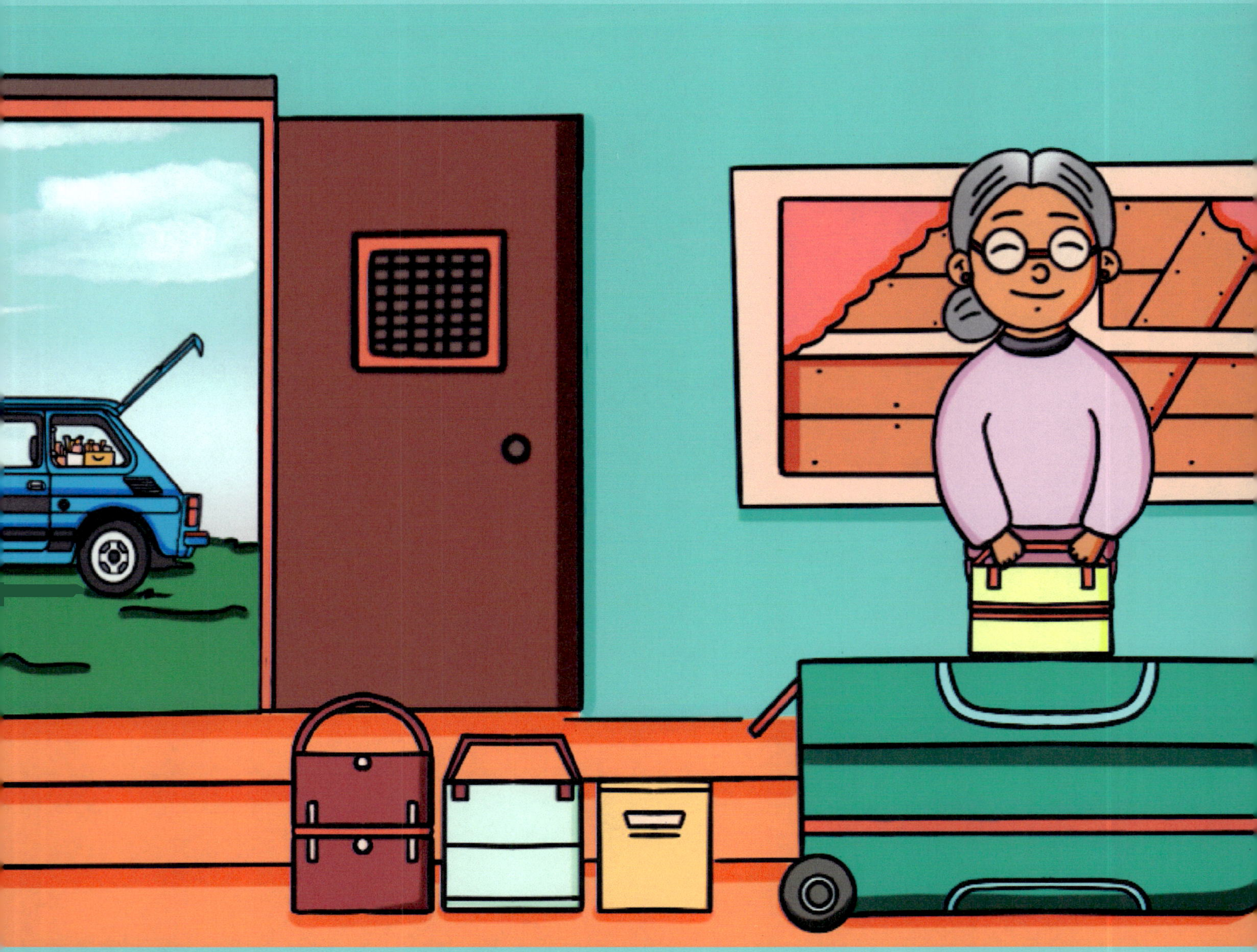

Lola sings,
“This is the way we load the car” as
the family packs the car.

"Let's go get the girls."

Dad says, “I guess they don’t KNOW what happens to chickens who don’t evacuate for a hurricane.”

"Why is he doing that?"
"Your dad is protecting our house from the wind by covering the windows and doors."

Lola follows her dad's example and prepares the chicken coop for the coming storm.

See you soon house!

As moms, we have had years of opportunities to use an opportunity as a teachable moment. This book is designed to help prepare so that you don't have to use an act of mother nature as a teachable moment.
- Kate & Carrie

If you enjoyed this book, please take a moment to review it on Amazon or tell your bookstore about it.

To say thank you for reading & sharing it, we would like to offer you a free coloring book. Just go to https://carrieecasey.clickfunnels.com/hurricane

Resources

- www.youtube.com/user/usweathergov
- www.weatherwizkids.com/weather-hurricane.htm#:~:text=A%20hurricane%20is%20a%20huge,contact%20with%20warm%20ocean%20waters
- www.ready.gov/hurricanes
- www.youtube.com/watch?v=vP2z8QAmQAQ
- www.Youtube.com/watch?v=pGdmODvMrE4
- www.natgeokids.com/za/discover/geography/physical-geography/hurricanes/

Hurricane Preparedness Supplies

- Reusable ice packs
- Portable, self-powered light source (candles, flashlights, lanterns)
- Gas or diesel container
- Batteries (rechargeable preferred)
- Bungee cords & tie-downs
- Flexible waterproof sheeting
- Important papers
- Baby supplies (food, diapers, ointments)
- Radio (battery, solar or hand-crank
- Coolers & ice chests
- Bottled water
- Canned food
- Can opener & utensils
- Medications
- Sanitary supplies
- Pets
- Pet food
- First aid kit
- Generator (optional)

Made in the USA
Monee, IL
05 November 2026

67380137R00021